NISTIR 6657

A Rational Methodology for Estimating the Luminous Flux Based upon Color Primaries from Digital Projection Displays

Edward F. Kelley
Optoelectronics Division
Electronics and Electrical Engineering Laboratory

Karl Lang
Lumita, Inc.

Louis D. Silverstein
VCD Sciences, Inc.

Michael H. Brill
Datacolor, Inc.

January 2009

U.S. Department of Commerce
Carlos M. Gutierrez, Secretary

National Institute of Standards and Technology
Patrick D. Gallagher, Deputy Director

A Rational Methodology for Estimating the Luminous Flux Based upon Color Primaries from Digital Projection Displays

Edward F. Kelley, Karl Lang , Louis D. Silverstein and Michael H. Brill

ABSTRACT: A standard methodology exists for estimating the flux from front projection displays by sampling the projected illuminance of a white source signal. With the advent and use of white projection primaries, a dramatic increase in flux can be achieved over the combination of red, green, and blue primaries alone. However, when a white primary is used saturated-color areas in an image are constrained to low flux levels relative to the display maximum and further undergo a perceptual compression in relative lightness. As a result, bright saturated colors cannot be rendered accurately and the appearance of full-color imagery is distorted. The display of color-accurate imagery does not generally use the white primary due to these problems, and thus there is a need for providing an equivalent flux measurement that will better describe the performance of all projectors when rendering full-color imagery. Background research and a method utilizing the source-signal color primaries are described.

KEYWORDS: color accuracy; color output; color perception; color rendering; display measurement; display metrology; ICDM Display Measurements Standard; IEC 61947-1; ISO 21118; light output; front projector, flux measurement, projection display; sampled luminous flux measurement; white primary; white subpixel

1. INTRODUCTION

It has been observed that digital projection-display devices that include a white (W) image field or white sub-pixels in addition to the "standard" set of R (red), G (green) and B (blue) primaries can reduce the effective perceptual color gamut of projected information. [1, 2] This reduction is manifest by intended bright, saturated colors being constrained to low flux values, both absolutely and relative to the maximum white of the display. In such RGBW displays, the white "primary" darkens bright colors in two ways: (1) Relative to RGB displays, the illuminance of pure R, G, or B is low because the white primary must share the screen area or active projection time even when it is shut off, and (2) The visual system tends to adapt to the prevailing light so that perceived white stays the same; hence a high-illuminance white will depress the perceived brightness of other colors. To see adaptation working in the reverse direction, one can make bright, saturated colors look even brighter and more saturated by artificially depressing the white in a scene. [3] Such variations in the effective perceptual color gamut of imaging devices such as projection displays may be readily characterized via perceptual experiments with human observers and can be estimated by the use of three-dimensional color spaces such as CIELAB or color appearance models such as CIECAM02. [4, 5] However, rather than address the subtleties of visual adaptation and its effects on perceived color, we choose a metric for RGBW displays that is based on simple optics, namely the sum of the illuminances of the full-on pure primaries R, G, and B.

Whereas the color gamut reductions described above may not be important for the presentation of simple text and some graphics, a reduction in the perceptual color gamut can be highly objectionable when imagery is viewed. Thus, we consider here two modes of projector operation: a **text/graphics mode** and an **imagery mode**. The text/graphics mode can use a white primary component in addition to the RGB primaries. The imagery mode will use only the RGB primaries without any contribution from the white primary. We limit our discussion to front projectors that are intended to use standardized RGB signals such as the sRGB specification. [6] We want to be able to characterize the projected luminous flux of any projector in both modes. The use of the term "pixel" refers to the smallest division of the screen that can produce all colors. Whereas we are limiting this investigation to front-projector flux estimations, the thinking presented here can be extended to luminance measurements of both rear-projection displays and emissive displays that include white primaries or white subpixels.

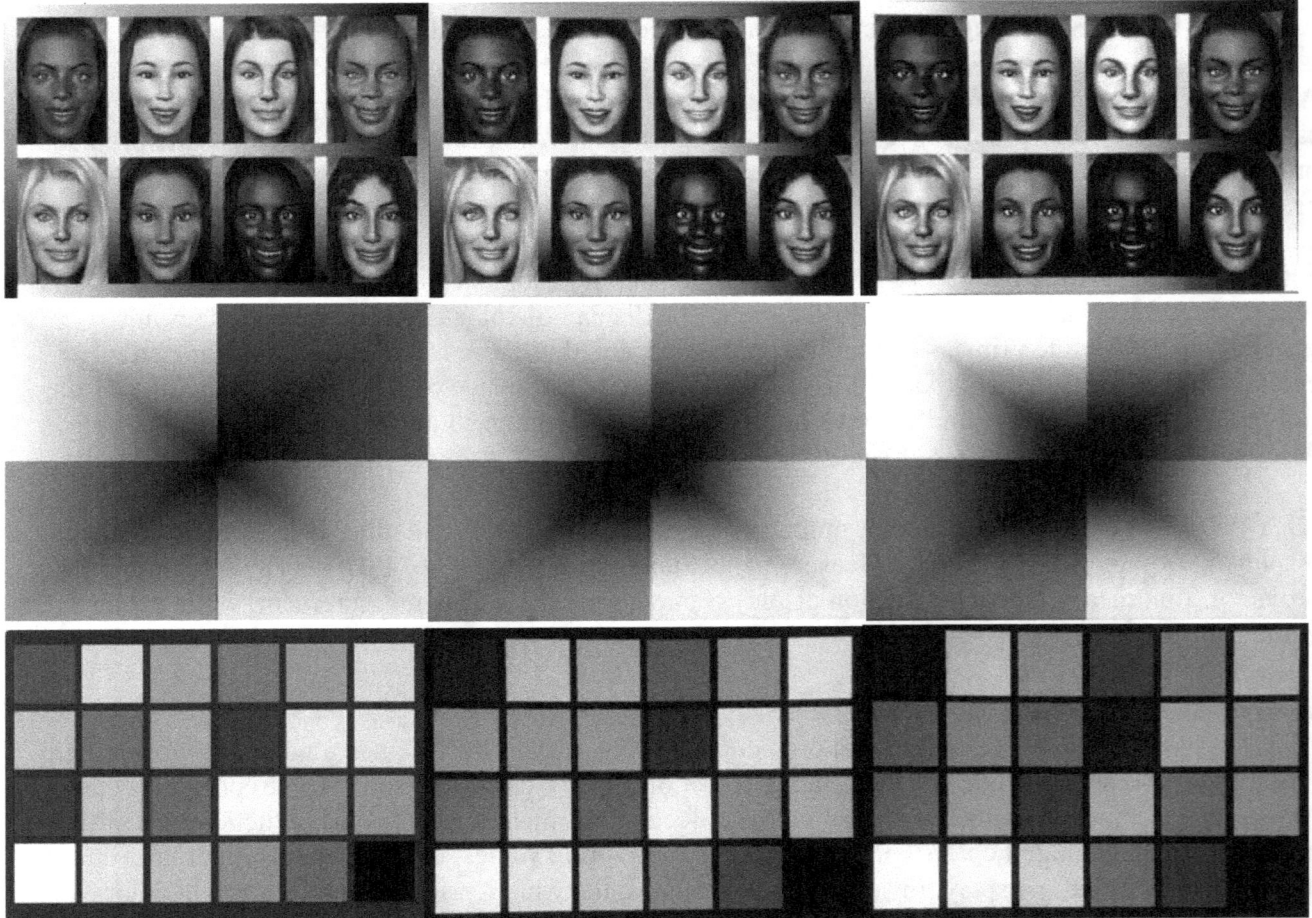

Figure 1. Color rendering with and without a white primary (**left**, original digital image; **middle**, photograph of screen with RGB projector without a white primary; **right**, photograph of screen with RGBW projector that includes the white primary).

Figure 1 shows an example of a face pattern [7], a multi-colored ramp, and a test pattern based upon the Macbeth ColorChecker® [8] that depict the difference between the text/graphics mode and imagery mode in a projector that includes a white primary component. The left image is the original digital image sent to the projector. The middle image is a photograph of the screen with the projector in the imagery (RGB) mode, and the right image is a photograph of the screen with the projector in the text/graphics mode including a white primary component (RGBW). You will note that the introduction of the white primary tends to provoke a relative darkening of the colors compared to the white level along with shifts in the perceived hue and color saturation. These photographs can provide only approximations to how the projected images actually look, but they convey a reasonable impression of how the white primary influences imagery.

To characterize the effective light output of digital projectors to be used for the accurate rendering of color imagery, and to compare the color-balanced performance modes of RGB projectors with those employing an additional white primary component, a method has been proposed in the literature by one of the authors that employs a nonatile (nine-tile) trisequence of RGB blocks—see Figure 2. [9] This method, which avoids the contributions from any white primary components, is compared with the full-white-screen method outlined in the existing measurement standards for fixed-resolution projectors. [8] Rather than measuring a white screen, this method measures the three tiled patterns and combines the results to provide a flux estimate based only upon the RGB primaries without a white primary contribution. Why not just use a white screen for a projector's imagery mode? To be sure that there is no white primary component present, the RGB patterns are used. Projectors often have several different modes of operation, and if they have a white primary present, the amount of contribution of the white primary can vary, depending upon the mode selected. For example, one digital projector has a mode called "sRGB" that also includes contributions from a white primary, whereas the sRGB specification does not account for any white primary component. The use of patterns consisting of only simultaneous RGB primaries on the screen is a safe method to assure that no confounding white primary component is present. The patterns are referred to by their diagonal color: NTSR refers to nonatile trisequence pattern with a red diagonal, NTSG, to green, and NTSB, to blue.

Figure 2. Nonatile trisequence patterns for estimating projector flux without the contribution of white primary components. The rectangles are equal in dimensions ±2 px, and in each pattern, the nine tiles fill the entire screen. From left to right the patterns are designated by their diagonal color from upper left to bottom right, NTSR, NTSG, and NTSG, respectively.

The following sections discuss the theory and apparatus used in making the measurements and the results of those measurements. The conclusion contains our recommendations. The appendix contains an example of measurement entries for the Display Measurements Standard document that is currently being developed by the International Committee for Display Metrology (ICDM).

2. RADIOMETRY AND PHOTOMETRY THEORY

Please note: Throughout this discussion, we speak of a measurement of the luminous flux (or spectral radiant flux) of a front-projector. Although we will usually refer to this as simply flux, it is to be understood that it is, in reality, a sampled illuminance (or spectral irradiance) measurement of the projected image multiplied by the area of the projected image to obtain an estimation of the luminous (or spectral radiant) flux. We will employ the correct terminology "flux" whereas the projection industry and existing or proposed standards use the coined term "light output" for measurements of the white screen and "color output" for measurements of the flux that use the nonatile trisequence patterns (Figure 2) where any white primary component and other complications are avoided. For reference, Table 1 lists the variables and symbols used in this document.

In part, the method employed to measure the flux is based upon an existing standard. [10] That standard calls for a sampled measurement of the illuminance E_{Wij} of a projected white image in nine places at the center of rectangles that divides the full screen approximately (±2 px) into a 3×3 matrix of equally sized rectangles—see Figure 3. The estimated flux of the white screen Φ_W is the product of the average illuminance and the measured area A of the image;

Table 1. Variables or symbols reserved for use in this document (and possible future extensions)

Variable or Symbol	Definition
D = R, G, B, C, M, Y, K, or W	displayed color = red, green, blue, cyan, magenta, yellow, black, or white
λ	wavelength of visible light (our measurement range is 380 nm to 780 nm)
Φ, I, L, E, M	luminous flux Φ (in lumens, lm), luminous intensity I (in candela, cd = lm/sr), luminance L (cd/m^2), illuminance E (lux = lm/m^2), luminous exitance M (lm/m^2)
$\Phi(\lambda), I(\lambda), L(\lambda), E(\lambda), M(\lambda)$	spectral radiant flux (W/nm), spectral radiant intensity (W/sr/nm), spectral radiance (W/sr/m^2/nm), irradiance (W/m^2/nm), radiant exitance (W/m^2/nm)
$\Phi_D, \Phi_D(\lambda)$	luminous flux of color D, spectral radiant flux of color D
$E_D, E_D(\lambda)$	illuminance of color D, spectral irradiance of color D
$L_D, L_D(\lambda)$	luminance of color D, spectral radiance of color D
A, A_{box}	area of projected image, area of 1/3 screen-size box
Φ_{RGB}	estimated luminous flux based upon a nine-point sampled illuminance measurement and a measurement of the area A of the projected image for which only the primary RGB pixels are employed with no white primary
Φ_W	estimated luminous flux based upon a nine-point sampled illuminance measurement and a measurement of the area A of the projected image for which a white primary may be included with the RGB primary pixels in a white screen
$E_{ij}, E_{ij}(\lambda), E_{Dij}, E_{Dij}(\lambda)$	illuminance or spectral irradiance in general (E_{ij}) or of color D (E_{Dij}) at any of the nine sample points with row $i = 1, 2, 3$ and column $j = 1, 2, 3$, where we use the standard matrix orientation with E_{D11} denoting the upper left, E_{D13} the upper right, E_{D31} the lower left, E_{D33} the lower right, etc. Summations indicated by the summation symbol with indices ranging from $i, j = 1, 2, 3$ to 3 will represent the usual mathematical arrangement where each index ranges from 1 to 3 covering the entire matrix: $$\sum_{i,j=1}^{3} x_{ij} = x_{11} + x_{12} + x_{13} + x_{21} + x_{22} + x_{23} + x_{31} + x_{32} + x_{33},$$ where x represents any quantity.
px	abbreviation for pixel

$$\Phi_W = \frac{A}{9} \sum_{i,j=1}^{3} E_{Wij}. \tag{1}$$

The illuminance (spectral radiance) measurement is made with the illuminance (spectral radiance) meter detector head in the focal plane of the projected image, and the illuminance meter must be cosine corrected. In the existing standard there are no position accuracy statements for the location of the illuminance measurements. Further, there is no specification for the calculation of the area when it is not perfectly rectangular, nor is it specified how accurate the cosine correction must be or even how accurately the illuminance meter must be calibrated. We will add the following requirements:

1. Position accuracy of illuminance measurement: The illuminance (irradiance) measurement region must be positioned within a circle of diameter 2.5 % of the minimum of the screen height or width that is centered at each of the 3 × 3 rectangles evenly dividing the screen.
2. Cosine correction accuracy of illuminance (irradiance) meter: The uncertainty of the deviation of the cosine correction from ideal must be no greater than ±2 % over the measurement region of the centers of the 3 × 3 rectangles evenly dividing the screen.
3. Illuminance (irradiance) meter accuracy: The relative uncertainty with a coverage factor of two of the illuminance (irradiance) meter must be 4 % or less and traceable to a national metrology institute (NMI). [11]

All the measurements made for this document confined the position accuracy to within 2 % or less of the centers of the 3 × 3 rectangles.

The area A is the area of the projected image on the projection screen (a plane in space and not an actual screen for all the measurement is made for this document). Because it is difficult to position the projected area exactly onto a specific size of screen, we must account for the actual projected size of the image. Given a general planar convex quadrilateral where the diagonals are specified by vectors $\mathbf{p}$ and $\mathbf{q}$—see Figure 4—the area is given by half the magnitude of the cross product of the diagonal vectors: [12]

$$A = \frac{1}{2}|\mathbf{p} \times \mathbf{q}| = \frac{1}{2}\begin{Vmatrix} \mathbf{e}_x & \mathbf{e}_y & \mathbf{e}_z \\ p_x & p_y & p_z \\ q_x & q_y & q_z \end{Vmatrix} = \frac{1}{2}\left|(p_y q_z - p_z q_y)\mathbf{e}_x - (p_x q_z - p_z q_x)\mathbf{e}_y + (p_x q_y - p_y q_x)\mathbf{e}_z\right|. \tag{2}$$

Assuming that the measurement plane is the x–y plane, so that all components in the z–direction are zero, this becomes

$$A = \frac{1}{2}|\mathbf{p} \times \mathbf{q}| = \frac{1}{2}\left|(p_x q_y - p_y q_x)\right|, \tag{3}$$

where x is the horizontal position (positive to the right) and y is the vertical position (positive up) with the origin at the bottom left of the measurement plane defining the imaginary screen in our case. To establish a simple estimation of the uncertainty in the area measurement, define a square with the same area as this distorted rectangle $A = s^2$. Given that each side has the same uncertainty δs in its measurement, to a good approximation the relative uncertainty of the area measurement is

$$\delta A / A = \sqrt{2}\, \delta s / s. \tag{4}$$

At the time the original standards were written and adopted with revisions by other standards organizations, white primaries (or, in fact, any additional primaries used in combination with R, G and B) were not commonly present in projection systems. Because of the different modes that can include the contributions of a white primary, we make the distinction between the projector flux Φ_W with a possible white primary included and the projector flux Φ_{RGB} that does not include a white primary.

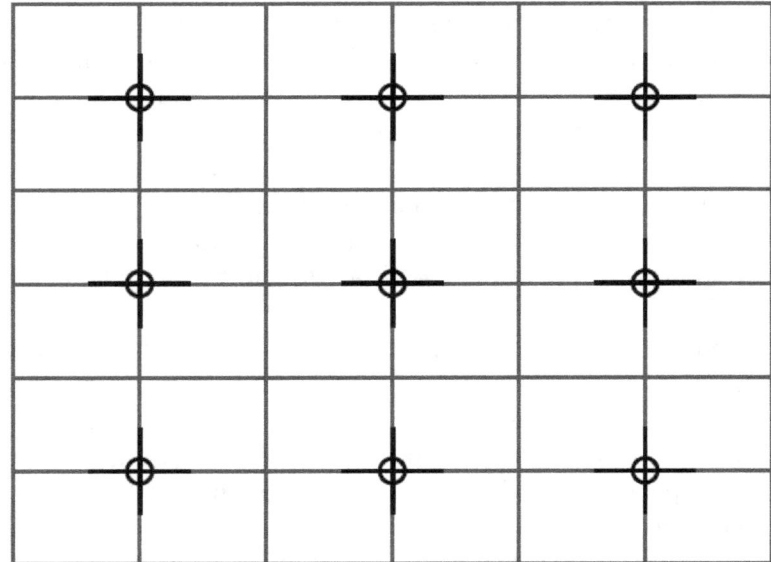

Figure 3. Patterns employed to establish measurement locations.

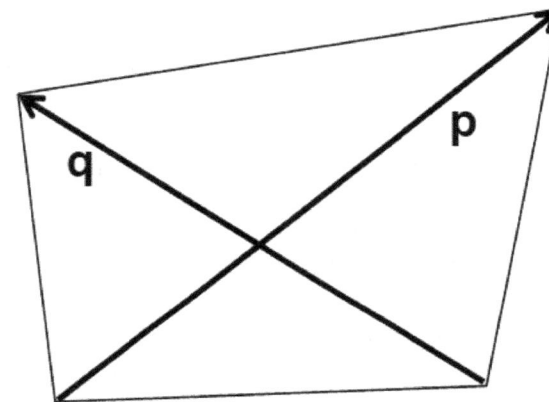

Figure 4. Area of projected image derived from diagonal vectors.

There is a setup pattern specified in the existing standard [6], the intension of which is to allow the operator to adjust the controls of the projector to capture a specified range of gray shades. The objection to this pattern is that it has the dark gray rectangles embedded within a white background. Even for a young eye, this makes it difficult to see the dark grays owing to veiling glare in the eyes (or disability glare as we get older). We have employed a similar pattern (named SET01S50) that uses a 50 % (gray level 127) background instead of white. It also includes 32-step ramps at the top and bottom—see Figure 5. [13]. A similar pattern SET01K (not shown here) has a black background that permits even better visibility of the dark levels. The actual pattern employed to check the projector setup for this document is shown in the bottom Figure 5. For most of the projectors measured it is possible to see all 32 levels in a user-configured mode. Often the white region is compressed for other modes so that not all levels are visible nearest the white level.

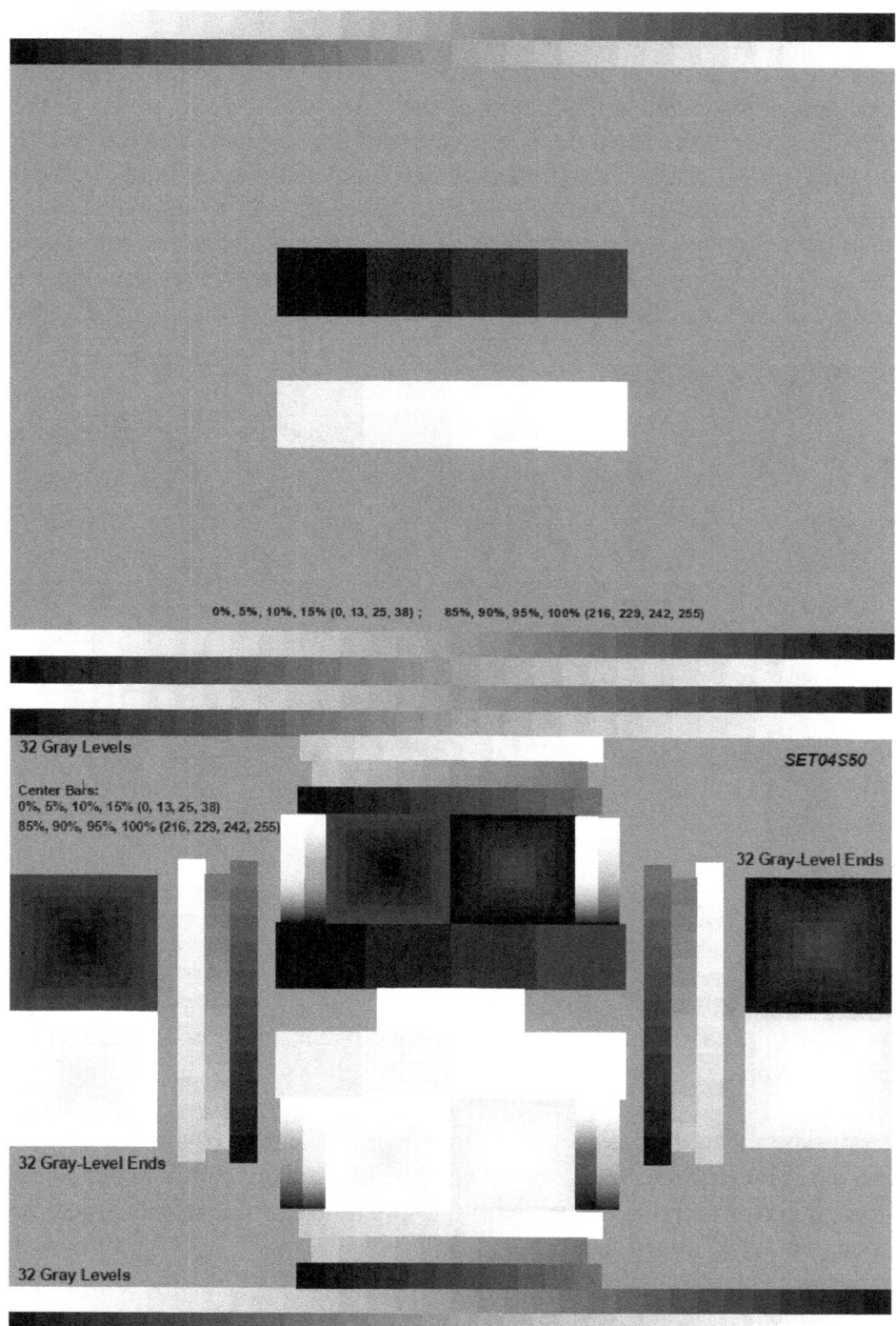

Figure 5. Patterns employed to establish proper setup of the projector. The top pattern (SET01S50) is a modified version of the pattern specified in the existing standard document. [6] The bottom pattern (SET04S50) is the pattern employed in setting up the projectors for this document.

3. MEASUREMENT APPARATUS

The chosen size of the projected image is 1.2 m × 0.9 m, having an aspect ratio of 4:3. The area is approximately $A = 1.08$ m^2. A metal framework outlines the intended image area with black felt behind and above the framework to reduce contamination from scattered light. The projector rests on rails that are orthogonal to the image plane and at the bottom of the framework. In the corners of the framework are millimeter arrays that permit the accurate measurement of the projected image—see Figure 6. The uncertainty in the framework is estimated to be 2 mm. We estimate the uncertainty in the horizontal or vertical measurement of the projected image to be 6 mm to account for any imprecision in the sharpness of the focus of the projected image at the edges as well as accuracy in measurement of the placement of the image. Referring to Eq. (4), $s = 1.039$ m, and we will take $\delta s = 6$ mm. The resulting relative component uncertainty in the area of the projected image is estimated to be $\delta A/A = 0.82$ %.

Figure 6. Laboratory setup. The projectors are placed on a rail system for positioning. The video generator is outside of the picture at the left.

Figure 7 shows the spectral irradiance measurement of the center of the last of the nonatile trisequence patterns (NTSB). The spectroradiometer-integrating-sphere combination is positioned via x–y linear stages programmed to move the center of the sphere to the nine measurement points. The apparatus is operated manually. The procedure is as follows:

1. The projector is placed on the platform, turned on, and fed a VGA (video graphics adapter) white full screen.
2. The voltage levels of the signals are checked to be preferably within ±1 % of the desired 0.7 V for the VGA signal.
3. Various modes of operation of the projector are investigated to provide the brightest mode and, if the projector includes a white primary, the mode that avoids the white primary.
4. The first, brightest, mode is selected.
5. The projector is positioned to approximately fill the framework that defines the screen area, and the corners of the projected image are measured to calculate the actual area A of the projected image.

6. The mode is viewed with a high-speed, low-resolution spectrometer to document the operation of the selected mode—whether it contains a white-primary contribution or not. (By this time, the projector has fully warmed up.)
7. The grayscale and greenscale for the projector is then quickly measured with a simple luminance meter and a white plaque on the face of the integrating sphere at the center of the white or green image. This documents the electro-optical transfer function (affectionately often called a "gamma" curve) for the projection mode for a gray screen and for a saturated green screen—see Figure 9.
8. The full-screen white and black is then measured and a full-screen (sequential) contrast is calculated with corrections for stray light, where the corrections are obtained with the aid of a projection mask (discussed below).
9. The spectral irradiance (illuminance) of a white full screen is measured at the nine locations to provide a flux Φ_W for a white screen.
10. The spectral irradiance (illuminance) is measured at the nine locations for each of the nonatile trisequence patterns, and a flux Φ_{RGB} is calculated avoiding any white-primary contributions.
11. The next mode of operation is selected and steps 5 through 10 are repeated. Each mode that is selected for investigation receives the same documentation.

For measurements using three nonatiling trisequence patterns, the equivalent illuminance E_{ij} for any location i, j is a combination of the illuminances from each pattern at that location (see Figure 8):

$$E_{ij} = E_{Rij} + E_{Gij} + E_{Bij}. \tag{5}$$

The estimation of the flux Φ_{RGB} is obtained by multiplying the average equivalent illuminance for the nine locations by the projected area:

$$\Phi_{RGB} = \frac{A}{9} \sum_{i,j=1}^{3} E_{ij} = \frac{A}{9} \sum_{i,j=1}^{3} (E_{Rij} + E_{Gij} + E_{Bij}). \tag{6}$$

The flux Φ_{RGB} should include no contribution from a white primary. For measurements using the white screen, the flux Φ_W,

$$\Phi_W = \frac{A}{9} \sum_{i,j=1}^{3} E_{Wij}, \tag{7}$$

may include contributions from a white primary, depending upon the mode of the projector.

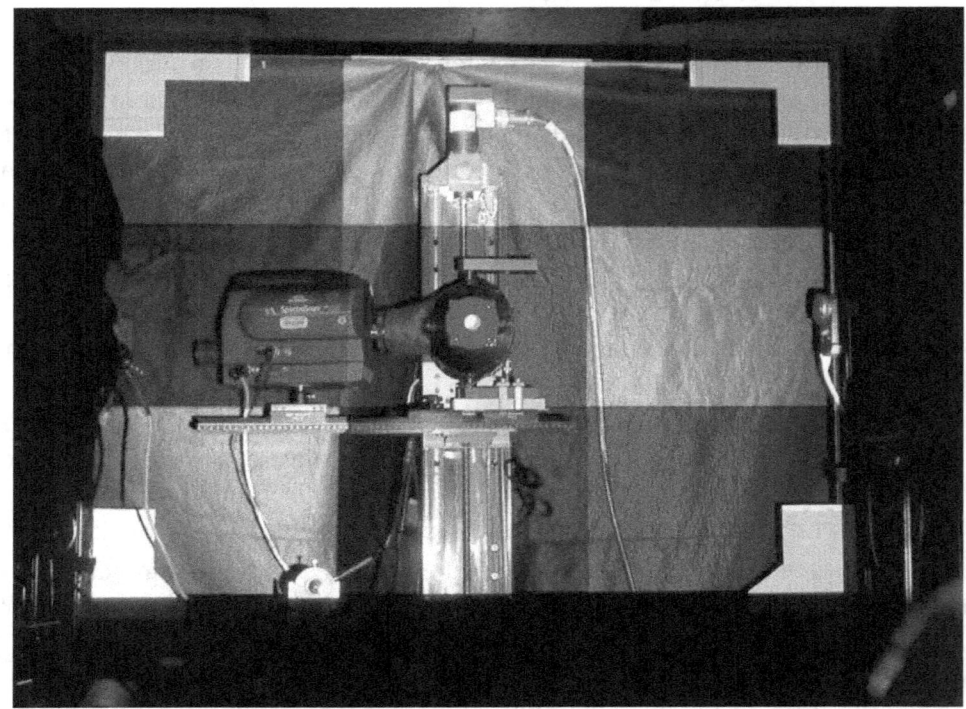

Figure 7. Measurement of the third nonatile trisequence pattern.

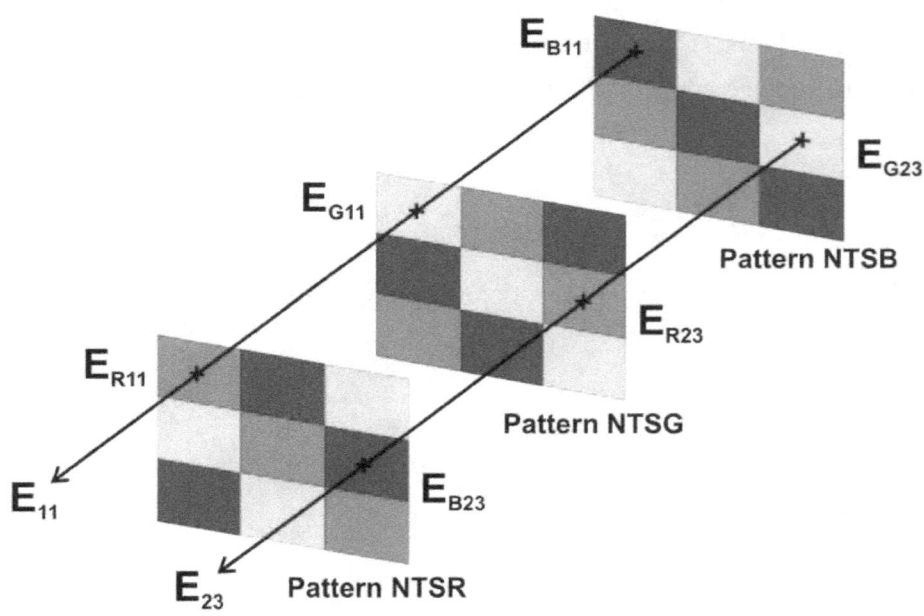

Figure 8. Use of nonatile trisequence patterns to establish illuminances, shown for two locations.

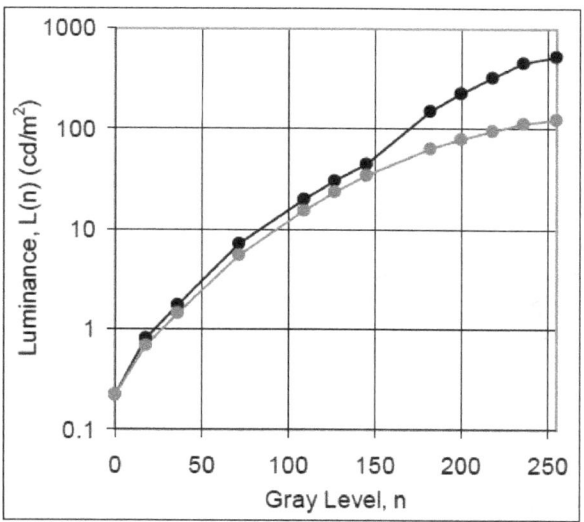

Figure 9. Electro-optical transfer function ("gamma" curve) for a projector with a white primary. The upper curve is for a white screen, and the lower curve is for a saturated green screen.

3.1.1 DETERMINATION OF COSINE RESPONSE

For these measurements, it is critical that the integrating-sphere detector exhibit a proper cosine correction over the range of angles for which it will be used. Figure 10 shows a horizontal cross-section of the spectroradiometer peering into the integrating sphere and measuring the wall radiance on the opposite side toward the front of the sphere. A photopic photodiode is implanted in the wall of the sphere to provide a monitor to complement the spectroradiometer measurement result. The thick interior wall of the detector is made from sintered high-purity powdered polytetrafluoroethylene (PTFE). (The photopic photodiode is not cut through to the interior wall so that the diffusing properties of the wall material help make the photodiode more Lambertian in its response.) The left side of Figure 11 shows an EKE lamp mounted on a rotary stage with

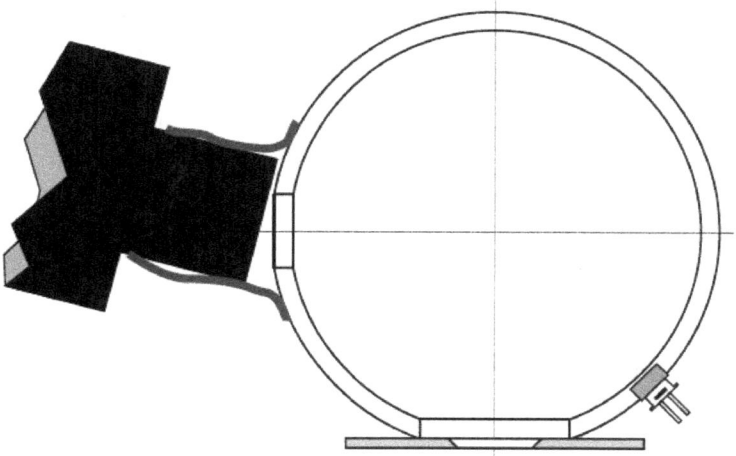

Figure 10. Horizontal cross-section of integrating-sphere detector. A photopic photodiode is embedded in the wall without destroying the interior surface. When the spectroradiometer is used, a black-felt multilayer wrap surrounds the lens and near surface of the sphere to avoid any introduction of stray light.

its rotation axis directly beneath the entrance port of the integrating-sphere detector. (A color meter is used instead of a spectroradiometer for this measurement only. The color meter was used to test various types of cosine detectors available to make this measurement.) The right side of Figure **11** shows the resulting normalized data as a function of the angle of the source from the normal of the detector. This provides us with an estimate of the relative uncertainty in the cosine correction of $u_{cc} = 0.38$ %, rendering the detector suitable for use over $\pm 30°$ that covers the angles for the nine measurement locations. The horizontal scan that is used here represents the worst-case uncertainty. Previous measurements show that a vertical scan with a horizontal detection (as is used here) shows much less uncertainty.

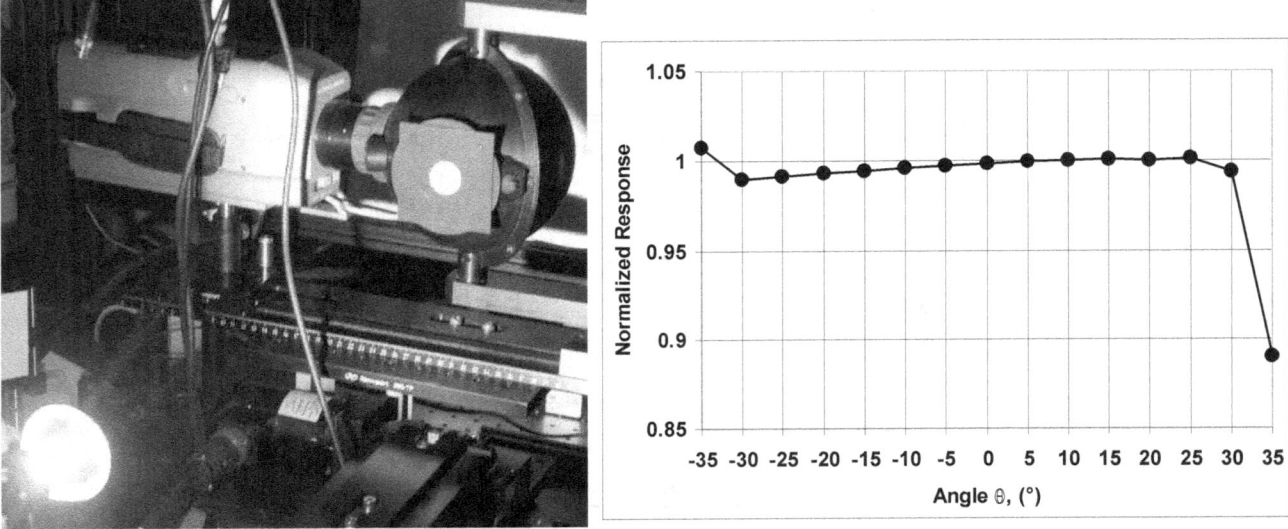

Figure 11. Apparatus using EKE lamp to determine the cosine-correction capabilities of the integrating-sphere detector. A color meter is shown rather than the spectroradiometer that is used for the primary measurements.

3.1.2 CALIBRATION OF INTEGRATING SPHERE DETECTOR

A 150 mm diameter integrating-sphere source with a $D_s = (51.06 \pm 0.12)$ mm diameter exit port is used to calibrate the integrating-sphere detector to provide measurements of spectral-irradiance and illuminance. Figure 12 shows the spectroradiometer measuring the radiance (luminance) of the exit port of the integrating-sphere source (left picture). That source is then used to calibrate the integrating-sphere detector with the spectroradiometer employed to measure the wall radiance (middle and right pictures in Figure 12). Three luminance meters are also employed to measure the center of the integrating-sphere source. The luminance-meter results are compared with the spectroradiometer results for each measurement-field-angle settings ($0.125°$, $0.25°$, $0.5°$, $1°$, and $2°$). The three luminance meters agree with a standard deviation of 0.29 %. The five spectroradiometer measurement-field-angle settings agree with a standard deviation of 0.19 %. However, the deviation between the luminance meter averages and the spectroradiometer luminance averages was $u_{\text{dev}} = 1.9$ %.

Figure 12. Apparatus used to provide an illuminance calibration of the integrating-sphere detector. Left picture shows the spectroradiometer measuring the exit port of the integrating-sphere source; middle and right pictures show the source being used to calibrate the integrating-sphere detector with its spectroradiometer.

The illuminance E at the entrance port of the integrating-sphere detector can be calculated from a knowledge of the luminance L_{cs} of the calibration source, the sizes of the ports involved, and the distance z between the exit port of the source and the entrance port of the detector:

12

$$E = \frac{L_{cs}A_s}{z^2 + R_s{}^2 + R_d{}^2}, \qquad (8)$$

where A_s is the area of the source, R_s is the radius of the exit port of the source, and R_d is the radius of the entrance port of the detector; $A_s = \pi R_s{}^2$. During the calibration, it is not practical to reposition the apparatus to directly measure the luminance of the source. Rather, we rely on the source-monitoring photopic photodiode. The photopic photodiode on the integrating-sphere source is calibrated based upon luminance measurements over the exit-port area made by one of the luminance meters. The calibration of this source photopic photodiode is $f_m = 578.4$ (cd/m^2)/μA with a relative standard deviation of $u_m = 0.23\ \%$ (this uncertainty doesn't include the uncertainty in the luminance measurement). The luminance L_{cs} of the integrating-sphere source for calibrating the detector is then given by

$$L_{cs} = f_m\,J_{spd}, \qquad (9)$$

where J_{spd} is the source photopic photodiode current. The integrating-sphere source is placed approximately seven exit-port diameters from the detector port. The illuminance calculation represented by Eq. (8) is independently checked against an illuminance-meter measurement of the illuminance at the integrating-sphere-detector port and found to be within 1 % of the measured value.

We anticipated receiving a spectrally calibrated source to check all the meters used in this calibration effort. However, the calibration source was damaged in shipment and could not be used. The individual luminance meters are considered to have a relative combined standard uncertainty of $u_{lm} = 2\ \%$ in measuring a tungsten-halogen illuminant. The relative combined standard uncertainty of the spectroradiometer making a similar measurement is estimated to be $u_{sr} = 2\ \%$. In order to compensate for the loss of the calibrated source, we will accept u_{sr} as the spectroradiometer component uncertainty, but we will add in the deviation u_{dev} and treat it as another independent component uncertainty. Small changes in position and focus of the spectroradiometer had no influence in its wall-luminance (radiance) measurement result when used to measure the wall radiance of the integrating-sphere detector. In fact, the detector configuration is remarkably robust. Several times the detector sphere housing was slammed into the framework when an error occurred in the positioning while the apparatus was being configured; each time the detector was re-configured and re-calibrated. Through all of this setting up and re-configuration the calibration of the integrating-sphere detector survived at $f_d = 13.76$ lx/(cd/m^2) with a relative standard deviation of only $u_d = 0.47\ \%$. When in use, the luminance L_s of the interior wall of the integrating-sphere detector is measured. The illuminance E is given by

$$E = f_d\,L_s. \qquad (10)$$

The factor f_d is good for either photometric or radiometric quantities.

3.1.3 MEASUREMENT OF SCATTERED LIGHT

As shown in Figure 13, a projection mask is employed to estimate the stray-light contribution to the measurement result. [14] The stray-light is measured for all nine positions, averaged, and the relative standard deviation is calculated to be $u_{sl} = 0.13\ \%$. The computer screen is on during the course of these stray-light measurements as it is for the illuminance measurements in the nine locations.

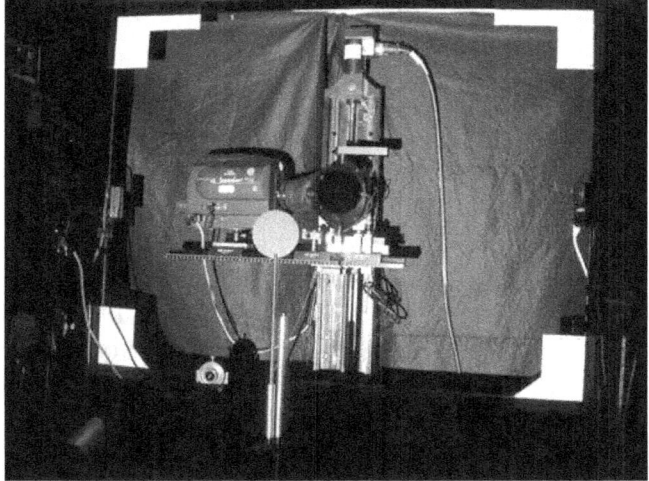

Figure 13. A projection mask is used to determine stray-light contributions.

3.1.4 UNCERTAINTY

We combine the above component uncertainties to obtain the relative combined standard uncertainty and relative expanded uncertainty with a coverage factor of two in Table 2. The type of uncertainty listed in the third column has two classifications, roughly speaking, Type A is an uncertainty established by statistical means (a random effect in the current measurement process), and Type B is an uncertainty based upon other evaluations (a systematic effect in the current measurement process)—see reference [11] for more details. We will treat the deviation between the luminance meters and spectroradiometer when measuring the center of the integrating-sphere source as an added component uncertainty in our measurement apparatus—as explained above.

In general, a flux measurement Φ based upon an illuminance E measurement involves the area A illuminated:

$$\Phi = E\,A. \tag{11}$$

The relative uncertainty $\delta\Phi/\Phi$ in the flux measurement involves the uncertainties in the illuminance E measurement and the measurement of the screen area A:

$$\left(\frac{\delta\Phi}{\Phi}\right)^2 = \left(\frac{\delta E}{E}\right)^2 + \left(\frac{\delta A}{A}\right)^2. \tag{12}$$

From Eq. (8), the relative uncertainty in the illuminance measurement arises from its calibration

$$\left(\frac{\delta E}{E}\right)^2 = \left(\frac{\delta L_{cs}}{L_{cs}}\right)^2 + \left(\frac{\delta A_s}{A_s}\right)^2 + u_{geo}{}^2, \tag{13}$$

where

$$u_{geo}{}^2 = 4\left[\frac{z^2\delta z^2 + R_s{}^2\delta R_s{}^2 + R_d{}^2\delta R_d{}^2}{(z^2 + R_s{}^2 + R_d{}^2)^2}\right] \tag{14}$$

is a geometry component uncertainty, L_{cs} is the luminance of the calibration source, $A_s = \pi R_s{}^2$ is the area of the exit port of the calibration source with a radius of R_s ($R_s = 25.53$ mm, $\delta R_s = 0.03$ mm; $\delta A_s/A_s = 2\delta R_s/R_s = 0.23$ %), z is the distance from the calibration source to the detector entrance port ($z \cong 355$ mm, $\delta z = 0.5$ mm), and R_d is the radius of the detector entrance port ($R_d = 12.74$ mm, $\delta R_d = 0.03$ mm). The geometric component relative uncertainty is $u_{geo} = 0.28$ %.

All the contributing uncertainties are straightforward, except for the luminance of the calibration source. From our discussions above, we can write

$$\left(\frac{\delta L_{cs}}{L_{cs}}\right)^2 = u_{cc}{}^2 + u_{sr}{}^2 + u_{dev}{}^2 + u_{sl}{}^2 + u_m{}^2 + u_d{}^2. \tag{15}$$

Here, $u_{cc} = 0.38$ % is the relative component uncertainty from the cosine correction, $u_{sr} = 2$ % is the uncertainty in the spectroradiometer, $u_{dev} = 1.9$ % is the deviation between the luminance meters and the spectroradiometer that we are adding in as another component uncertainty, $u_{sl} = 0.13$ % is the relative stray-light contribution being added in rather than serving as a correction, $u_m = 0.23$ % is the relative component uncertainty in the luminance of the calibration source based upon its photodiode output current, and $u_d = 0.47$ % is the reproducibility of calibration of the integrating-sphere detector. The result of this analysis is that the relative expanded uncertainty with a coverage factor of two is estimated to be 6 %.

Table 2. Relative uncertainties for flux calculation based upon sampled illuminance or irradiance measurements.

1	Cosine correction contribution (u_{cc}, no correction made for this)	Type B	0.38 %
2	Calibration of spectroradiometer (u_{sr})	Type B	2.0 %
3	Deviation between the luminance meter and spectroradiometer (u_{dev})	Type B	1.9 %
4	Scattered light contribution (u_{sl}, no correction made for this)	Type B	0.13 %
5	Contribution from calibration of photopic photodiode monitor used as a calibration lamp monitor (u_m)	Type A	0.23 %
6	Repeatability in illuminance inference from wall luminance measurements that use the spectroradiometer and 150 mm diameter integrating sphere (u_d)	Type A	0.47 %
7	Source area relative uncertainty contribution ($\delta A_s/A_s$)	Type A	0.23 %
8	Geometrical component relative uncertainty (u_{geo})	Type A	0.28 %
9	Projection area measurement relative uncertainty ($\delta A/A$)	Type A	0.82 %
	Relative combined standard uncertainty:		**3 %**
	Relative expanded uncertainty with a coverage factor of two:		**6 %**

3.1.5 DETERMINATION OF COMPOSITION OF WHITE AND PRIMARIES

A high-speed, low-resolution spectrometer is employed to determine how each projector renders a white or primary-color screen in each mode of operation. In order to be sure that a certain projection mode does not add in a white primary component, the spectra of the white screen can be monitored in time. A collimated detector consists of a short-focal-length lens with a polished end of a 2 mm diameter plastic fiber-optic cable with black protective sheath placed at its focal point. The cable is routed to the spectrometer. The end of the cable at the spectrometer is elongated vertically and narrowed horizontally. A lens collimates the output of the cable into a diffraction grating, and another larger lens focuses the spectral image onto the detector array of the high-speed camera—see Figure 14. Figure 15 shows the results from the high-speed spectrometer when a white screen is measured within one frame. Figure 16 shows the spectra as measured with the spectroradiometer having a measurement-field angle of 2° resulting in a full-width-at-half-maximum (FWHM) bandwidth of 20 nm.

Figure 14. High-speed low-resolution spectrometer. The left image shows the collimated detector located at the base of the projector image. The right image shows the spectrometer with, from left to right, fiber input, collimating lens, grating, imaging lens, and high-speed camera.

Figure 15. High-speed low-resolution spectrometer rendering of R, G, B, and W for a projector with a white primary. The R spectrum is made brighter to make it more visible. These spectra are obtained within one displayed frame of a white screen.

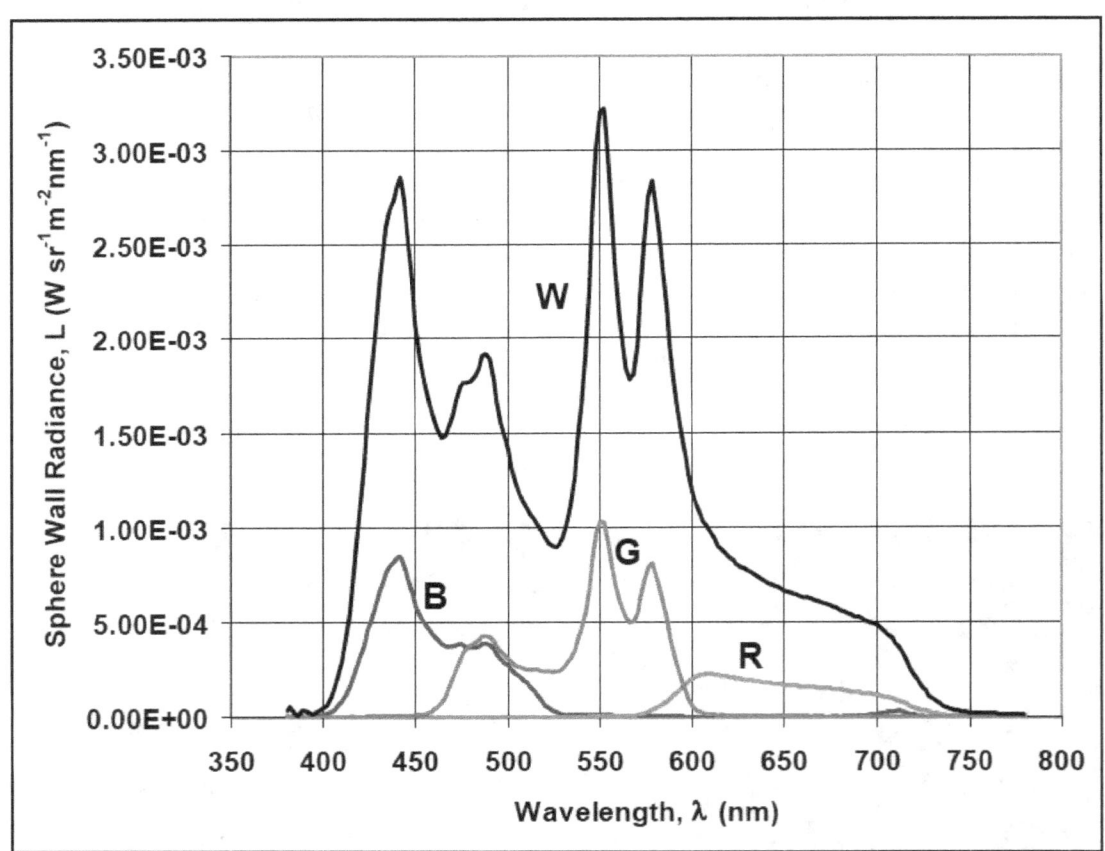

Figure 16. Comparison of integrating-sphere wall spectral radiance for R, G, B, and W for a projector with a white primary. These spectra are accumulations of many frames of R, G, B, and W screens respectively. A 2° measurement-field angle is used in the spectroradiometer, giving it a bandwidth of 20 nm FWHM.

4. MEASUREMENT RESULTS

Five front projectors were measured; three included a white primary and two did not—see Table 3. The projectors not containing a white primary measured approximately the same for a white screen as for the nonatiling trisequence. The projectors containing a white primary showed dramatic increases in flux from the nonatiling trisequence to the white screen, as much as a 70 % flux increase. The expanded relative uncertainty with a coverage factor of two is estimated to be 6 % for these flux measurements.

Table 3. Luminous flux Φ based upon sampled illuminance measurements.

Front Projector	White Primary Available? (Y/N)	Φ_W White Screen (lm)	Φ_{RGB} RGB Nonatiling Trisequence (lm)	RGB Color Contribution to Flux (%)	White Primary Contribution to Flux (%)
A	Y	1471	464	32%	68%
B	N	1063	1067	100%	NA
C	Y	1894	578	31%	69%
D	Y	1937	588	30%	70%
E	N	2027	2060	100%	NA
					NA = not applicable

5. CONCLUSION

We have seen how the introduction of a white primary component in digital front-projection displays may well produce a large luminous flux ("light output") based upon a white full screen. However, when the white primary component is eliminated for the purposes of the display of accurate color image information, the resulting flux ("color output") can be significantly reduced with only the RGB primaries being used. A method is needed to ensure that no white primary component is employed in the total luminous flux measurement of digital front-projection display modes used for the presentation of full-color imagery. This will help ensure that the luminous flux measurements are properly related to the available, effective perceptual color gamut of the display in color-critical imagery modes, as well as provide a more meaningful specification of projection display flux (light output) in these modes for both manufacturers and consumers. The luminous flux estimation based upon the illuminance measured from the nonatile trisequence patterns are found to be a meaningful and reliable means of characterizing the flux available in color-critical modes from all digital projectors, including those that use a white primary component to enhance their flux (light output).

6. APPENDIX: EXAMPLE OF MEASUREMENT PROCEDURE ENTRY

In this section we provide examples of various measurement methods as they might be contributed to the upcoming International Committee for Display Metrology (ICDM) document, Display Measurement Standard (DMS). It is important to note that the philosophy of the ICDM DMS is that the included measurements represent a buffet of measurements where we can pick and choose as we wish; no specific measurements are required. Section references and pattern references in these examples are to sections and patterns in the DMS document.

6.1   FRONT PROJECTOR MEASUREMENTS — INTRODUCTION

We consider the measurement of front projectors and front-projection screens in this section. Often, illuminance measurements are performed on such projectors. We speak of illuminance measurements throughout this section, whereas we could equivalently consider irradiance measurements instead—the reasoning is the same.

6.1.1 DETECTOR UNCERTAINTY:

Detectors used (often illuminance meters) should have a relative expanded uncertainty of 4 % or less that is traceable to a national metrology institute. The deviation of relative spectral responsivity from the spectral luminous efficiency of the human eye for photopic vision (f_1') must be 8 % or less.

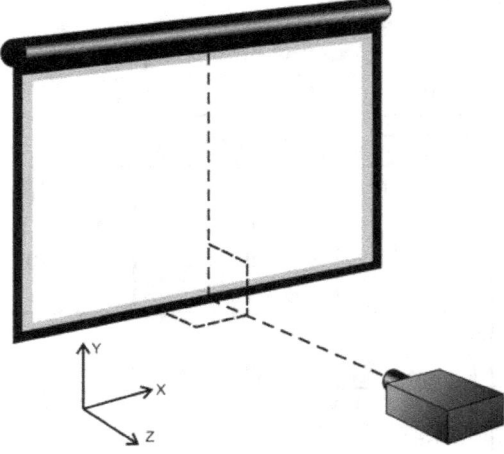

Figure 1. Projector placement.

6.1.2 DARKROOM REQUIREMENTS:

Even with black walls, a black screen, and careful control of any instrument lights (including computers), a projection room is not completely dark. Reflections of light off the screen (even if black) will bounce off the walls and apparatus to contaminate the illumination falling on the screen from the projector. Thus, rather than strictly requiring stray-light illumination to be below some minimum, we can attempt to make corrections for stray light as needed. The use of a black screen with stray-light corrections can even permit the use of a room with white walls.

6.1.3 PROJECTOR PLACEMENT:

The placement of the projector relative to the screen should be detailed in the manufacturer's specifications. Often the lens axis of the projector will be orthogonal to the vertical line at the center of the screen and placed either at a level near the bottom or above the screen. The image plane is usually vertical, parallel to the x-y plane. The projector is usually placed on or attached to a horizontal surface parallel to the x-z plane. See Figure 1.

6.1.4 VIRTUAL SCREEN:

A virtual screen is a vertical plane in space where the projected image would be focused if a projection screen were placed in that plane. Instruments to measure the light are often placed within and behind the framework so that their detector inputs are along the image plane. See Figure 2.

One way to provide a virtual screen is to construct a black framework to define a surface with black material provided behind the framework to reduce scattered light into the room. The face of the framework defines the virtual screen surface. Millimeter grids can be accurately placed in the corners of the framework in such a way as to permit an accurate measurement of the location of the corners of the projected image. The desired accuracy of the placement of the grids to define the projected image in this way is 0.02 % or less of the minimum of the horizontal and vertical size of the projected image; for a projected area of 1.333 m × 1 m this requires a grid placement accuracy of 2 mm or less.

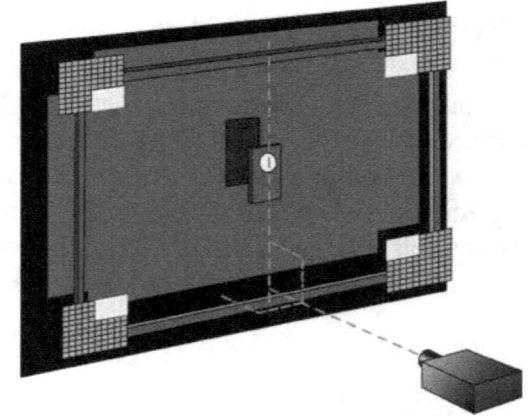

Figure 2. Virtual screen with black background.

6.1.5 PROJECTION MASK:

Projection masks permit the determination of the stray-light corruption of projection measurements made in darkened rooms. A projection mask is a thin matte-black disk from 1.5 to 3 times the diameter of the acceptance area diameter of the detector. The projection mask is used to shadow the detector from the direct rays from the projector and is placed from 30 cm to 60 cm in front of the detector—the larger projection masks being place at a greater distance from the detector. With the projection mask in place, the detector output is a measure of the stray-light contamination from the room and can be different for each pattern displayed. Black screens and a darkened room are preferred. See Figure 3. If a black screen is not readily available, then a darkened room will suffice.

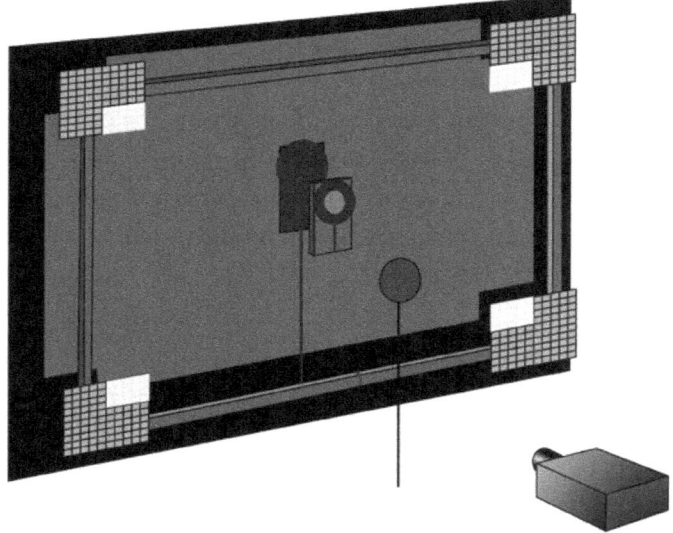

Figure 3. Projection mask in darkened room for stray-light measurement.

However, the projection mask has been found not to work particularly well in bright rooms such as bright conference rooms.

6.1.6 SLET—STRAY LIGHT ELIMINATION TUBE:

Stray-light-elimination tubes (SLETs) permit the measurement of the projected illuminance by rejecting stray-light corruption even in well-lit rooms—see Figure 4. One version of a SLET is constructed with five frusta: Four are in pairs back-to-back, and one is at the end to prevent light scattering off the illuminance meter from reflecting back to the illuminance meter and corrupting the measurement. The entry frustum has a slightly smaller inner diameter than the next three frusta so that the light from the projector doesn't illuminate the inner diameters of the second set of frusta—if at all possible. The interior of the tube and the frusta are all gloss black. The idea is to control the stray light to virtual extinction by multiple reflections rather than trying to diffusely absorb it. For clarity the interior of the tube is not shown in the bottom of Figure 4.

If the SLET is constructed so that the illuminance meter must be tilted in the

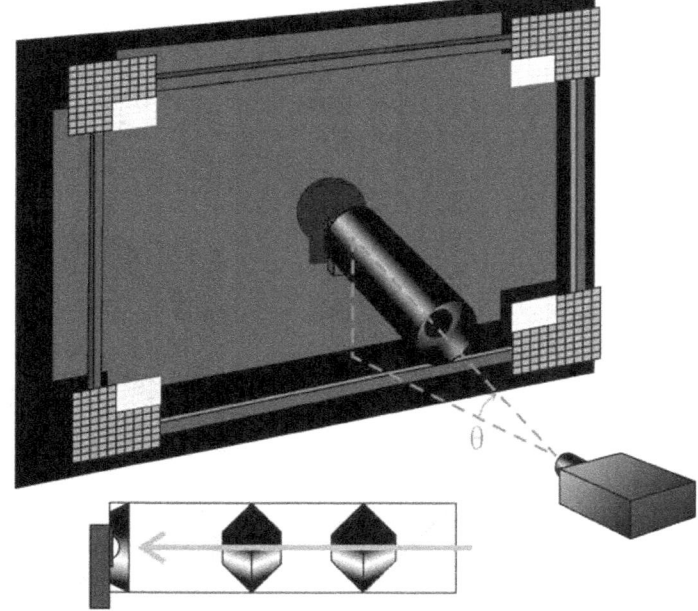

Figure. 4. Stray-light-elimination tube (SLET) with a cutaway showing interior frusta.

direction of the projector so that the illuminance meter is flat against the back of the SLET, then an angular correction must be made to the resulting measurement, E_{SLET}, if it is required that the illuminance measurement, E, be made parallel with the image plane:

$$E = E_{SLET}\cos\theta, \qquad (1)$$

where θ is the angle from the normal of the image plane. Simpler versions of the SLET offer only three or even two interior frusta along the tube with a corresponding possible increase in the admission of stray light. By sighting up the SLET from the illuminance meter position using your eye, it is possible to inspect for stray-light entering the SLET. Judicious placement of the frusta can virtually eliminate the stray light from the room.

6.1.7 WHITE REFLECTANCE STANDARDS USED TO INFER ILLUMINANCE:

White reflectance standards (we will refer to them as pucks here) are often made from sintering powdered material into a disk shape. Often their hemispherical diffuse reflectances are from $\rho = 0.98$ to over 0.99. Rather than the use of an illuminance meter, some have placed these pucks in the image plane and measured their luminance L in order to determine the illuminance E via

$$E = \frac{\pi L}{\rho}. \tag{1}$$

Strictly speaking this is *not* correct. The use of a diffuse reflectance value of, say, $\rho = 0.99$ in Eq. (1) is true *only* for uniform hemispherical illumination. In general, it is *not* correct to use this relationship for the luminance meter and projector at various angles from the normal of the puck. These pucks are not perfectly Lambertian, as the use of Eq. (1) with the hemispherical reflectance would suppose.

To use such a puck to determine the illuminance, the puck would have to be calibrated for the geometrical configuration in which it is used. For projection systems the reflectance factor $R(\theta_s, \phi_s, \theta_d, \phi_d)$ would be required, where the source (projector) is at angles (θ_s, ϕ_s) relative to the normal of the puck and the detector (luminance meter) is at angles (θ_d, ϕ_d) relative to the normal—see Figure 5. Changing any of those angles can significantly change the value of the reflectance factor. Therefore, the correct relationship is

$$E(\theta_s, \phi_s) = \frac{\pi L(\theta_s, \phi_s, \theta_d, \phi_d)}{R(\theta_s, \phi_s, \theta_d, \phi_d)}, \tag{2}$$

where the reflectance factor calibration properly accounts for the source and detector angles employed. Depending upon the angles used, the error in by use of Eq. (1) can be as much as 10 % or larger.

For measuring full-screen contrasts with the luminance meter, the projector, and the puck at the same location for each measurement, the pucks can be useful. Under such full-screen-contrast measurement conditions, the expression for the contras,t

$$C = \frac{L_W}{L_K}, \tag{3}$$

holds true, where L_W and L_K are the luminance measurements using the puck at center screen. This assumes that the stray light in the room comes only from back reflections from the screen illumination and not from various sources of stray light, such as instrumentation lights or computer screens.

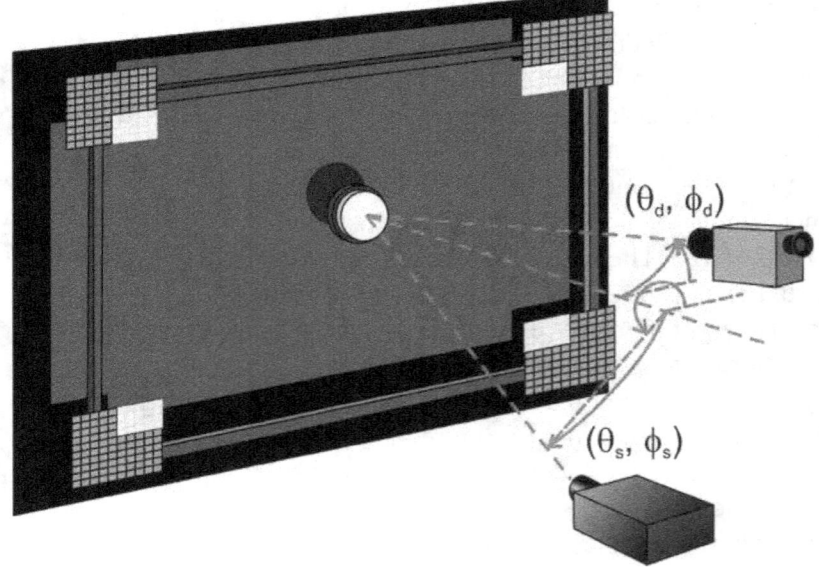

Figure 5. White reflectance standard used to infer illuminance.

6.1.8 PATTERN GENERATION:

It is strongly recommended that a calibrated video generator be employed whenever patterns are to be measured. If a computer output is employed, then the signal to the projector must be measured to assure it is correct—this is particularly true when analog signals are used. Any deviation from the appropriate signal standard employed must be noted. Pattern resolution (pixel array) from the generator should match the native resolution of the projector. Patterns are named and defined in Section ZZZ.

6.1.9 DETECTOR COSINE CORRECTION:

To measure illuminance in the projection plane, the illuminance meter must be kept parallel to the measurement plane and must be cosine corrected to within 1 % over the range of angles between the projection lens center and the measurement locations that are used. If this requirement is not met, then the cosine-correction error must be reported. Cosine correction means that for uniform illumination from a distant source, the illuminance E changes as

$$E = E_0 \cos\theta, \tag{4}$$

where E_0 is the illuminance with the source in the normal direction of the illuminance meter, and θ is the angle of the source from the normal.

6.2 AREA OF FRONT PROJECTOR SCREEN IMAGE

DESCRIPTION: We measure the rectangular or quadrilateral area of a projected image from a front projector displaying a white pattern on a front-projection screen. **Unit:** m^2. **Symbol:** A.

APPLICATION: Front projectors.

SETUP: The following icons are defined in Section 3.7 for any standard setup details:

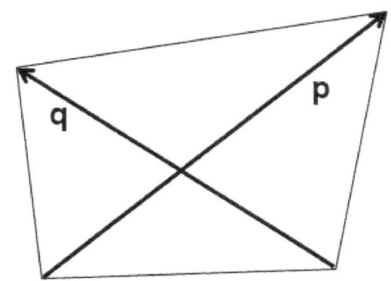

Figure 1. Diagonals of a planar convex quadrilateral.

OTHER SETUP CONDITIONS: **(1)** The front-projection screen used (virtual or real) must provide a means to determine the location of the corners of the projected image to within a distance of 0.02 % of the minimum of the horizontal or vertical projected image size. **(2)** A native-resolution full-screen white pattern is required (pattern FW*.* or equivalent). **(3)** It is preferable that the projected image have an area of no less than 1 m^2. **(4)** Some projectors provide trapezoidal adjustments, but the image does not have to be perfectly rectangular to determine its area. A quality video generator is strongly recommended.

PROCEDURE: The procedure depends on whether or not the projected image is rectangular:
1. **Rectangular Image:** If the projected white image corners define a rectangle, then make a straightforward measurement of the horizontal dimension H and vertical dimension V of the image.
2. **Nonrectangular Image:** If the projected white image is *not* sufficiently rectangular, then determine the horizontal (p_x, q_x) and vertical components (p_y, q_y) of the diagonals of the projected white image (Figure 1): $\mathbf{p} = p_x\mathbf{e_x} + p_y\mathbf{e_y}$, and $\mathbf{q} = q_x\mathbf{e_x} + q_y\mathbf{e_y}$, where $\mathbf{e_x}$ and $\mathbf{e_y}$ are unit vectors in the horizontal and vertical direction, respectively. The measurement will require an accurate grid to locate the corners of the projected image. (If a virtual screen is used, then grid plates must be accurately placed in the corners of the framework.) Determine the (x, y) coordinates of the corners of the projected image. Our notation will be: Lower left is (x_{LL}, y_{LL}), lower right is (x_{LR}, y_{LR}), upper left is (x_{UL}, y_{UL}), and upper right is (x_{UR}, y_{UR}).

ANALYSIS: If the projected white image is rectangular, then the area of the screen is given simply by the product

$$A = H\,V. \tag{1}$$

If the projected white image is not rectangular, then the components of the diagonals are given by

$$p_x = x_{UR} - x_{LL}, \quad p_y = y_{UR} - y_{LL}, \tag{2}$$
$$q_x = x_{UL} - x_{LR}, \quad q_y = y_{UL} - y_{LR}.$$

Note that in Figure 1, q_x is negative. The area is then given by

$$A = \frac{1}{2}\left| \mathbf{p} \times \mathbf{q} \right| = \frac{1}{2}\left| p_x q_y - p_y q_x \right|. \tag{3}$$

REPORTING: Report the area in square meters as needed.

COMMENTS: This method assumes that the edges of the projected image are straight lines. See Section XXX for measurements of barrel and pincushion distortions. A reference for measuring a convex quadrilateral area is http://mathworld.wolfram.com/Quadrilateral.html.

—SAMPLE DATA ONLY— Do not use any values shown to represent expected results of your measurements.		
Nonrectangular Analysis and Reporting Example: (Virtual screen with corner grids with origin at the lower left corner)		
Corner	x (mm)	y (mm)
LL	-11	5
UR	1321	1000
LR	1307	13
UL	-18	997
$(p_x, p_y) =$	1.332 m	0.995 m
$(q_x, q_y) =$	-1.325 m	0.984 m
$A =$	1.315 m^2	

6.3 SAMPLED FLUX FROM PRIMARY RGB COLORS

ALIAS: Color Output

DESCRIPTION: We calculate the luminous flux from a front projector by use of sampled illuminance measurements of RGB primary colors and the area of the projected image so that the influence of white primaries (white image fields and/or white subpixels) are avoided. **Unit:** lumen (lm). **Symbol:** Φ_{RGB}.

APPLICATION: Front projectors.

SETUP: The following icons are defined in Section 3.7 for any standard setup details:

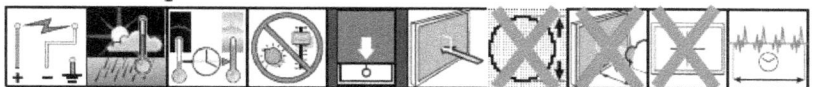

Figure 1. Pattern (SET01S50) to set up the front-projector.

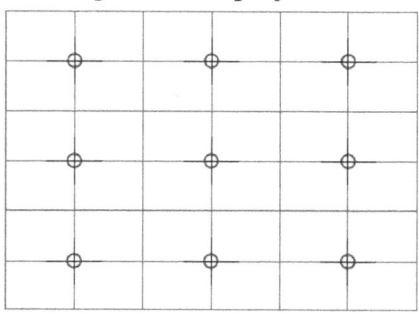

Figure 2. Pattern (AT02P) to determine detector positions.

OTHER SETUP CONDITIONS: The illuminance measurements must be made in a room where the stray light is ignorable or can be accounted for as a correction.

Patterns needed: **(1)** SET01S50 as in Figure 1, **(2)** AT02P to locate the centers of the 3x3 array of equal rectangles that defines the measurement grid (Figure 2), and **(3)** nonatile trisequence patterns (NTSR, NTSG, NTSB) consisting of three separate patterns with RGB 3x3 arrays (Figure 3). A video pattern generator is strongly recommended for the nonatile trisequence patterns in the native resolution of the projector.

Figure 3. Nonatile trisequence patterns (NTSR, NTSG, NTSB, respectively).

PROCEDURE: Select the mode of operation of the projector to be measured. For each mode selected perform the following steps:

3. Measure the area A of the projected image according to Section 6.2.
4. If possible, by adjustment of the projector settings, ensure that all the central dark gray and light gray levels in pattern SET01S50 (Figure 1) are discernable. Report any noncompliance.
5. Assure that the illuminance measurements are made at the nine centers of 3x3 equal (± 2 px) rectangles to within a radius of 2.5 % of the minimum of the screen height or width; this is the measurement grid. The use of a pattern as in Figure 2 (AT02P) can be helpful in determining the correct measurement locations.
6. Measure and record the illuminance of the three nonatile trisequence patterns (Figure 3) at the nine locations of the measurement grid.

ANALYSIS: We use the matrix notation i, j, where $ij = 11$ is the upper left and $ij = 33$ is the lower right, to define the locations of the measurement grid—see Figure 4, i = row, j = column. The illuminance E_{ij} at any location i, j is given by the sum of the contributions from each of the three patterns at that location:

$$E_{ij} = E_{Rij} + E_{Gij} + E_{Bij}. \tag{1}$$

The flux Φ_{RGB} is given by the product of the projected area and the average illuminance:

$$\Phi_{RGB} = AE_{ave} = \frac{A}{9} \sum_{i,\,j=1}^{3} E_{ij}. \qquad (2)$$

REPORTING: Report the flux Φ_{RGB} to no more than three significant figures (unless more can be justified by an uncertainty analysis).

COMMENTS: None.

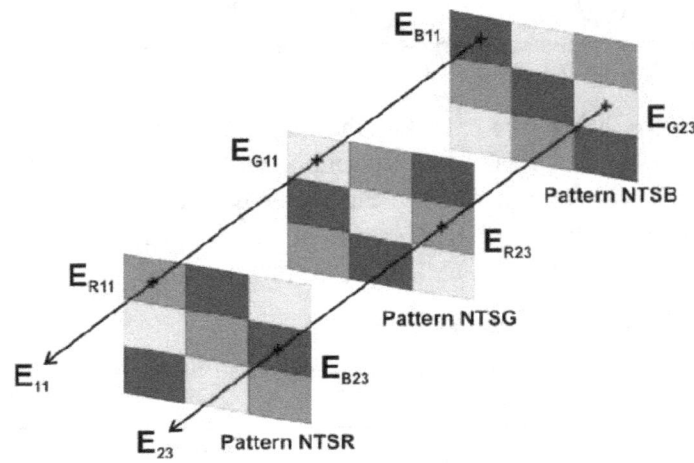

Figure 4. Illuminance measurements of nonatile-trisequence patterns.

Analysis Example:

Pattern	Illuminance, E (lx)		Illuminance, E (lx)		Illuminance, E (lx)	
NTSR	$E_{R11} =$	260.1	$E_{B12} =$	67.0	$E_{G13} =$	1319.6
	$E_{G21} =$	1521.9	$E_{R22} =$	323.6	$E_{B23} =$	65.8
	$E_{B31} =$	70.8	$E_{G32} =$	1618.2	$E_{R33} =$	320.7
NTSG	$E_{G11} =$	1409.0	$E_{R12} =$	301.5	$E_{B13} =$	61.9
	$E_{B21} =$	67.6	$E_{G22} =$	1578.3	$E_{R23} =$	318.7
	$E_{R31} =$	287.7	$E_{B32} =$	70.4	$E_{G33} =$	1455.8
NTSB	$E_{B11} =$	63.0	$E_{G12} =$	1459.9	$E_{R13} =$	289.2
	$E_{R21} =$	278.5	$E_{B22} =$	70.7	$E_{G23} =$	1407.6
	$E_{G31} =$	1543.9	$E_{R32} =$	333.5	$E_{B33} =$	63.6
Illuminance at each location:	$E_{11} =$	1732.0	$E_{12} =$	1828.4	$E_{13} =$	1670.7
	$E_{21} =$	1868.0	$E_{22} =$	1972.6	$E_{23} =$	1792.2
	$E_{31} =$	1902.4	$E_{32} =$	2022.2	$E_{33} =$	1840.1
Average:	$E_{ave} =$	1847.6	Area, $A =$	1.116	m^2 (Section 6.2)	
Flux $\Phi_{RGB} =$	2061.9	lm				

Reporting Example

Flux $\Phi_{RGB} =$	2060	lm (3 sig. figs.)

6.4 SAMPLED FLUX FROM WHITE

ALIAS: Light Output

DESCRIPTION: We calculate the luminous flux from a front projector by use of sampled illuminance measurements of a white full screen and the area of the projected image so that the influence of white primaries (white image fields and/or white subpixels) can be included. **Unit:** lumen (lm). **Symbol:** Φ_W.

APPLICATION: Front projectors.

SETUP: The following icons are defined in Section 3.7 for any standard setup details:

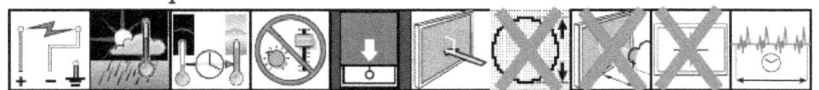

OTHER SETUP CONDITIONS: The illuminance measurements must be made in a room where the stray light is ignorable or can be accounted for as a correction.

Patterns needed: **(1)** SET01S50 as in Figure 1, **(2)** AT02P to locate the centers of the 3x3 array of equal rectangles that defines the measurement grid (Figure 2), **(3)** FW or a full-screen white pattern. A video generator that uses the native resolution of the projector is strongly recommended.

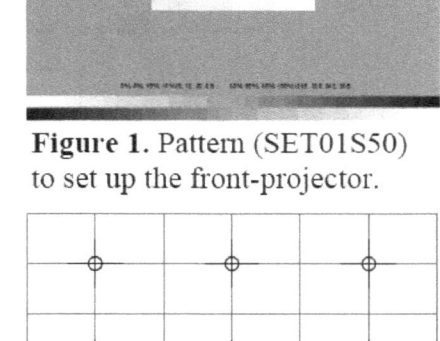

Figure 1. Pattern (SET01S50) to set up the front-projector.

Figure 2. Pattern (AT02P) to determine detector positions.

PROCEDURE: Select the mode of operation of the projector to be measured. For each mode selected perform the following steps:

1. Measure the area A of the projected image according to Section 6.2.
2. If possible, by adjustment of the projector settings, ensure that all the central dark gray and light gray levels in pattern SET01S50 (Figure 1) are discernable. Report any noncompliance.
3. Assure that the illuminance measurements are made at the nine centers of 3x3 equal (± 2 px) rectangles to within a radius of 2.5 % of the minimum of the screen height or width; this is the measurement grid. The use of a pattern as in Figure 2 (AT02P) can be helpful in determining the correct measurement locations.
4. Measure and record the illuminance of a full-screen-white projected image at the nine locations of the measurement grid.

ANALYSIS: We use the matrix notation i, j, where $ij = 11$ is the upper left and $ij = 33$ is the lower right, to define the locations of the measurement grid; i = row, j = column. The flux Φ_W is given by the product of the projected area and the average illuminance:

$$\Phi_W = AE_{ave} = \frac{A}{9} \sum_{i,j=1}^{3} E_{ij} . \tag{1}$$

REPORTING: Report the flux Φ_W to no more than three significant figures (unless more can be justified by an uncertainty analysis).

COMMENTS: This measurement method is an adaptation of *Electronic projection - Measurement and documentation of key performance criteria - Part 1: Fixed resolution projectors*, International Electrotechnical Commission, IEC 61947-1:2002(E), 40 pages, first edition 2002-08.

Analysis Example:

Illuminance at each location:	$E_{11} =$	1732.0	$E_{12} =$	1828.4	$E_{13} =$	1670.7
	$E_{21} =$	1868.0	$E_{22} =$	1972.6	$E_{23} =$	1792.2
	$E_{31} =$	1902.4	$E_{32} =$	2022.2	$E_{33} =$	1840.1
Average:	$E_{ave} =$	1847.6	Area, $A =$	1.116	m^2 (Section 6.2)	

Flux $\Phi_{WRGB} =$	2061.9	lm

Reporting Example

Flux $\Phi_{RGB} =$	2060	lm (3 sig. figs.)

7. REFERENCES

[1]   D R Wyble and M R Rosen, "Color management of DLP projectors," IS&T/SID 12th Color Imaging Conference, 228–232 (2004).

[2]   R. L. Heckaman and Mark D. Fairchild, "Effect of DLP projector white channel on perceptual gamut," Journal of the SID, Vol. 14, No. 9, pp. 755-761, (2006).

[3]   R. L. Heckaman and Mark D. Fairchild, "Expanding Display Color Gamut Beyond the Spectrum Locus," Color Research & Application, Vol. 31, No. 6, pp. 475-482, (2006).

[4]   Mark D. Fairchild, "Color Appearance Models, 2nd Edition," West Sussex, England: John Wiley & Sons Ltd, (2005).

[5]   CIE Technical Report 159 (2004): A colour appearance model for colour management systems: CIECAM02. Note that CIE is Commission Internationale de l'Eclairage (International Commission on Illumination).

[6]   IEC 61966-2-1 Ed. 1.0, Multimedia systems and equipment - Colour measurement and management - Part 2-1: Colour management - Default RGB colour space – sRGB.

[7]   E. F. Kelley, "Plotting the Course of the Next VESA Flat Panel Display Measurements Standard," J. Society for Information Display, Vol. 13, No. 1, pp. 67-79, January 2005.

[8]   Any commercial item referred to in this paper or presented in a photograph is identified only for the purpose of complete technical description. Such a reference does not imply recommendation or endorsement by the National Institute of Standards and Technology neither does it necessarily suggest suitability to task.

[9]   K. Lang, "Analysis of the proposed IEC 'Color Illuminance' digital projector specification metric, using CIELAB gamut volume," Late-Breaking-News Paper Presented at the 15th IS&T/SID Color Imaging Conference, Albuquerque, NM, November 2007.

[10]   *Electronic projection - Measurement and documentation of key performance criteria - Part 1: Fixed resolution projectors*, International Electrotechnical Commission, IEC 61947-1:2002(E), 40 pages, first edition 2002-08. This standard is an extension of older standards. The effort started with an ANSI (American National Standards Institute) standard [ANSI/NAPM IT7.215-1992, "Data Projection Equipment and Large Screen Data Displays--Test Methods and Performance Characteristics."]. A historical note: Because of that original standard, the estimated flux came to be termed incorrectly "ANSI lumens," a term that was not used in the document but became common around the industry. Such a term is *not* legitimate because ANSI is not the keeper of the lumen; hence, we discourage its use in any documentation or even informally. The measurement could properly be termed "ANSI flux" but *not* "ANSI lumen."

[11]   For a discussion of the proper way to specify uncertainty see: Barry N. Taylor and Chris E. Kuyatt, *Guidelines for Evaluating and Expressing the Uncertainty of NIST Measurement Results*, NIST Technical Note 1297, 1994 Edition.

[12]   http://mathworld.wolfram.com/Quadrilateral.html

[13]   E. F. Kelley and J. V. Miseli, "Setup Patterns for Display Measurements - Version 1.0," NISTIR 6758, 14 pp., June 2001.

[14]   P. A. Boynton and E. F. Kelley, "Compensation for Stray Light in Projection-Display Metrology," 2001-SID International Symposium Digest of Technical Papers, Society for Information Display, Vol. 32, pp. 334-337, San Jose, CA, June 4-8, 2001.